THE BANANA SLUG

A Close Look
at a Giant Forest Slug
of Western North America

Written and illustrated by
Alice Bryant Harper

Photographs by
Daniel Harper

Bay Leaves Press
Aptos California
1988

First Printing 1988
Second Print 1991, revised
Third Print 1993, revised
Fourth Print 1999

The four-color plates were prepared and printed by
Dai Nippon Printing Company, Ltd., Tokyo, Japan.

Cover by Mya Kramer Design

ISBN 0-9621218-0-0

Published by Bay Leaves Press
160 Robideaux Road
Aptos, California 95003 U.S.A.

in cooperation with
The Santa Cruz City Museum Association

CONTENTS

ACKNOWLEDGEMENTS

My thanks to those who advised and encouraged me in producing this book on the banana slug. My husband Dan has been willing to drop what he was doing and photograph banana slugs whenever I asked. He also edited the manuscript and gave needed suggestions. I am fortunate to have a partner so talented and willing to help.

Scott Harper generously shared his technical and teaching skills with his computer- illiterate mother. Jill Harper was willing to be the face on the book's cover and offered her support in many different ways.

Dr. Barry Roth and Dr. F. G. Hochberg edited and gave valuable scientific advice. Stan Stevens, Sandy Lydon, and Mort Marcus shared their knowledge of book publishing. I also appreciate the willing help of Hal Morris, Laurie Kiguchi, Joe Marciano, and Jim Goodhue.

The Museum Association and the museum staff have encouraged and assisted me from the onset of this project. I'm pleased to be associated with such a fine Natural History Museum and an exceptional group of people. John Lane, Tom McCarthy , Doug Petersen, and those on the Publications Committee were particularly helpful.

Love and curiosity of the natural world was passed on to me by my parents, Winifred and Sam Bryant. It has made all the difference.

Photo by Jill Harper.

The author and photographer with a slug from their garden.

PREFACE

It was over twenty years ago when I first became fascinated with the **banana slug**. My husband and I, along with our two small children, had just moved into a cabin tucked among California redwood trees in the Santa Cruz mountains. Since then, few days have passed without my seeing one or more of these slippery yellow creatures on our walkway or in our garden. I found their eggs, watched them grow from tiny hatchlings into elderly ten inchers, observed them climbing, burrowing, dropping via a slime cord, swimming, eating, being eaten, and witnessed their matings.

With growing curiosity, I began collecting all the information about this slug I could find, while keeping notes on my own observations. At that same time my husband, Dan, was becoming an avid nature photographer. He began photographing banana slugs doing ordinary and extraordinary things. We now have hundreds of slug photographs from throughout the animal's geographic range and file folders bulging with information. Giving lectures, organizing exhibits and writing articles about this (not exactly beloved) animal has earned me the dubious honor of being called the "Banana Slug Lady".

The size and color of the banana slug, along with its unusual mating habits, make it distinctive in the animal world. But most people who see this animal do not react with admiration or even fascination. They usually react with revulsion to its shape and texture. "Oh Yuck! What is that thing?" is a typical response. Because of such reactions, perhaps originating in the subconscious, many people think the banana slug unworthy of serious consideration. I tend to think of this as a human attitude problem, having little to do with the animal itself.

Our language even reflects our attitudes toward this slime-covered invertebrate. It is obviously not a compliment to be called a slug, sluggish, or slimy. Slug caricatures are usually shown in silly comical ways, whereas many other animals are anthropomorphically endowed with lofty human traits.

Courtesy of UCSC

Popular emblem at the University of California at Santa Cruz.

All this makes students at the University of California at Santa Cruz an exceptional group, when in 1986 they rebelled against their chancellor and demanded that the banana slug become their university mascot. A majority of the students took into account that the golden mollusk was a true native of their campus and more appropriate than the sea lion to decorate their sweatshirts and banners. Since then, the banana slug has also been proposed as California's official state mollusk.

In several northern California communities there are annual slug festivals. One such event features banana slug races. Each summer since 1968, California's Prairie Creek Redwoods State Park has The Banana Slug Derby. It is billed as "a celebration of this important old growth redwood creature".

Another festivity features a slug recipe contest with celebrities acting as tasters and judges. Banana slugs are gathered in the woods, confined a few days, killed, and cooked. The message seems to be - because this animal is locally prolific (and repulsive), it is dispensable. In my mind, it is not ecologically defensible to slaughter native, non-game animals just for laughs.

All this notoriety has brought a number of humorous commercial banana slug items to the marketplace, but hasn't significantly changed attitudes.

A banana slug is a wild animal and not by any stretch of the imagination suitable as a pet. Keeping one in a jar or terrarium without attention for even a short period of time could lead to its untimely death. Confinement of and experimentation on wild animals should be left to research scientists.

While the banana slug is particularly interesting to me, I would not claim it to be more or less worthy of admiration than any other wild animal in its native habitat. But, in an uncommonly beautiful part of the world, it is an integral link in the chain of life, and the acme of a long adaptive evolution. There is a beauty in its peaceful tenacious life and in its struggle for survival.

John Muir wrote, “When we try to pick out anything by itself, we find it hitched to everything in the universe". It is my hope that this very close examination of one small animal in the natural world will help illustrate the wonder and connection of all living things.

HABITAT

Foggy damp forests can be magical places. The shafts of sunlight break through the canopy of branches here and there to spotlight a lacy fern, a moss-covered rock near a stream or maybe a woodland slug nibbling on a mushroom. Such forests along the west coast of North America are the **habitat** for large slugs known as **banana slugs**. We will pluck this animal out of its niche and take a very close look.

Land slugs are pulmonate gastropod mollusks having an elongated body with no external shell. Mollusks are a group of invertebrate animals containing more than 60,000 living species, including slugs, snails, oysters, mussels, and squids. The phylum Mollusca is divided into six classes. Land slugs belong to the class Gastropoda (stomach-foot) and the subclass Pulmonata (with lung).

SPECIES DISTRIBUTION OF GENUS *ARIOLIMAX* IN CALIFORNIA

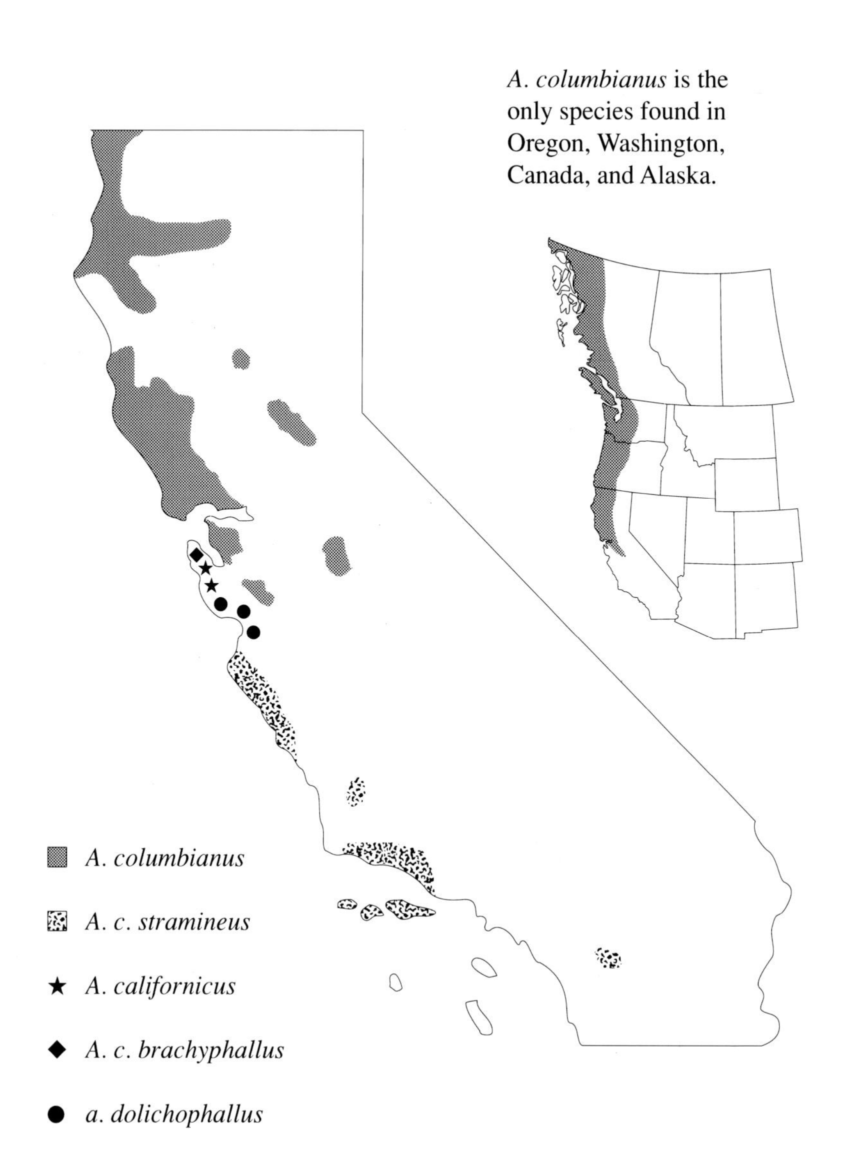

Map is based on research by Albert R. Mead (1943)

RANGE

The banana slug belongs in the genus *Ariolimax* and is indigenous to damp forests of western North America, including parts of the Coast ranges, the Olympic mountains, and several islands. It ranges from San Diego County of California in the south, northward to at least the southern part of Alaska. In 1991 the southern range was extended when an isolated population of banana slugs was documented at Palomar Mountain State Park in San Diego County of California. While it is found in California, Oregon, Washington, western Canada (including Vancouver Island) and Alaska, only California has all of the species and sub-species.

Within the genus are three species - *Ariolimax columbianus, Ariolimax dolichophallus, Ariolimax californicus* and two sub-species, *Ariolimax columbianus stramineus* and *Ariolimax californicus brachyphallus.* These species designations and their ranges are based on extensive research published in 1943 by Albert Mead. It was not external characteristics, such as their color or body size that caused the species to be separated. It was because of distinct internal differences, mainly in the genital structures.

The largest and most widespread group is *A. columbianus.* It is found as far north as Sitka, Alaska, west of the Cascade range in British Columbia, Washington, and Oregon; into California as far south as the Salinas Valley and on the western slopes of the Sierra Nevada in Tuolomne County. This is the only species that is sometimes spotted and has great color variations.

A. c. stramineus is separated from *A. columbianus* mainly by differences in the female structures. There are, however, more visible differences. *A. c. stramineus* is a smaller, thinner slug, it is unspotted, a yellow color, and has a smaller penis than *A. columbianus.* It is believed to be more primitive than the parent species. It is found in California from San Diego County northward to just below the Monterey Bay, including populations on Santa Cruz and Santa Rosa Islands.

The *A. dolichophallus, A. californicus,* and *A. c. brachyphallus* groups are restricted to a relatively small region between the Salinas Valley and the San Francisco Peninsula of California. They reflect a more recent evolution. Unlike *A. columbianus*, they are never spotted and are often a vivid butter yellow color.

It was in 1851 when the banana slug was first given a scientific name by Augustus A. Gould. Later, after more research and other populations were found, these names were changed and refined. The species names of animals and plants are in some way descriptive of that particular group. *Columbianus* is named after the Columbia River district where some of the first studies of this animal were done. *Stramineus* means straw-colored. *Dolichophallus* means long penis. *Californicus* is obviously named for the state of California and *brachyphallus* means short penis.

EVOLUTION

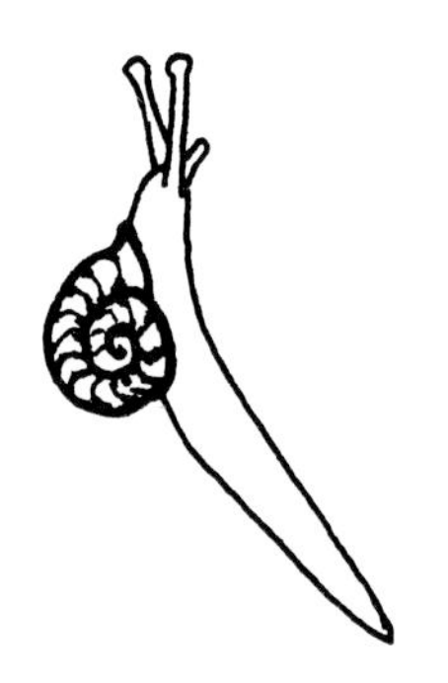

Slugs are believed to have evolved from snails and originally came from the sea. This is supported by the fact that the most primitive living mollusks are aquatic. When we look at existing freshwater, marine, or terrestrial gastropod mollusks, we can see living representatives of various stages of their evolution. There are land and aquatic snails with huge shells, others with progressively smaller shells, and so called semislugs that have only a tiny shell perched near their head or tail. From its life as a snail millions of years ago, the banana slug has the remnants of an internal shell plate remaining within its mantle head-covering.

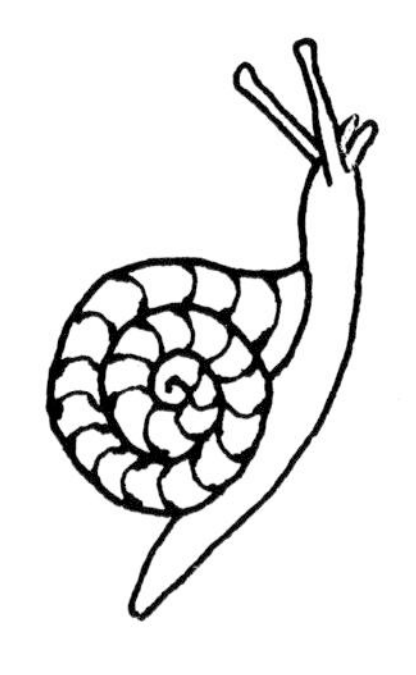

Anatomical studies of slugs show that the slug form has evolved from many different snail groups by parallel evolution both on land and in the sea. Therefore, slugs are not believed to be a group of closely related animals. Technically, we can only refer to "slug" as a type of body. This explains the enormous variety of slugs with differing life cycles, habits and body structures. It has been a difficult task for scientists to trace a particular slug back to its evolutionary beginning, because the trail backwards is so branched with many possible starting points. Also, few fossils exist of the slug's soft body parts to help us trace their evolution.

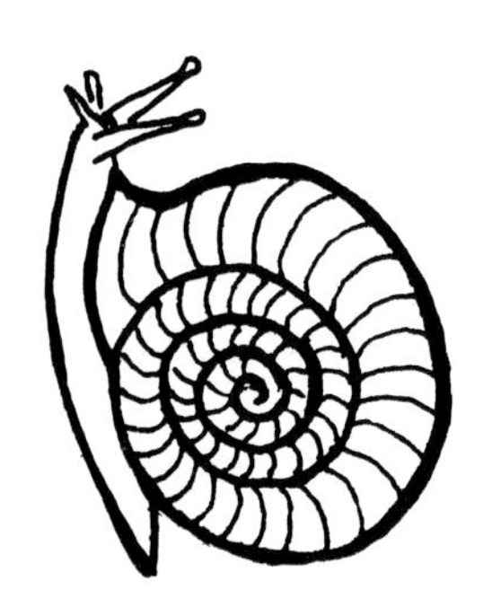

Slugs which have evolved onto land are limited to conditions of plentiful moisture, to areas with shelter from excessive light and heat, and tend to live where calcium is scarce. Snails, on the other hand, need a calcium source to build their shells. Slugs have eliminated that need. Land slugs lost a major protection from desiccation and from predators when they reduced or eliminated their shells, but they gained a wider range of environments. The flexible streamlined body, now free of the cumbersome shell, can crawl into small spaces and live underground for extended periods of time.

EGGS AND YOUNG

Banana slug eggs.

Banana slugs begin their lives as tiny colorless hatchlings only 15-20 mm (3/4 to one inch) in length. The parent has deposited a clutch of about 30 or more separate eggs in a hole or crevice where ground water and humidity are high. The whitish eggs measure about 7 mm (5/16 inch) and are nearly spherical. The thin shell can easily be flicked off to reveal a firm translucent egg containing the visible embryo. It takes three to eight weeks for the slugs to hatch. Hatching success is determined by sustained moisture, mild temperatures and the inability of predators to find the nest.

Immediately after hatching, the baby slugs are on their own, searching for food. As with many animals, it is at this early stage that they are most vulnerable to predators and the elements. A bird or shrew would find a tiny new banana slug to be a manageable meal. It is indeed the survival of the fittest—and the luckiest. The large number of eggs laid ensures that even with a high mortality rate, enough young will survive to continue the species.

The baby slug grows rapidly and in a few days begins to get its color or spots. There are differences of opinions as to how long a banana slug usually lives. Garden snails can live six years and the great gray slug in Europe can live three years. This gives us an idea of the probable life span of the banana slug.

Newly hatched colorless slugs with adult.

BODY PARTS OF A BANANA SLUG

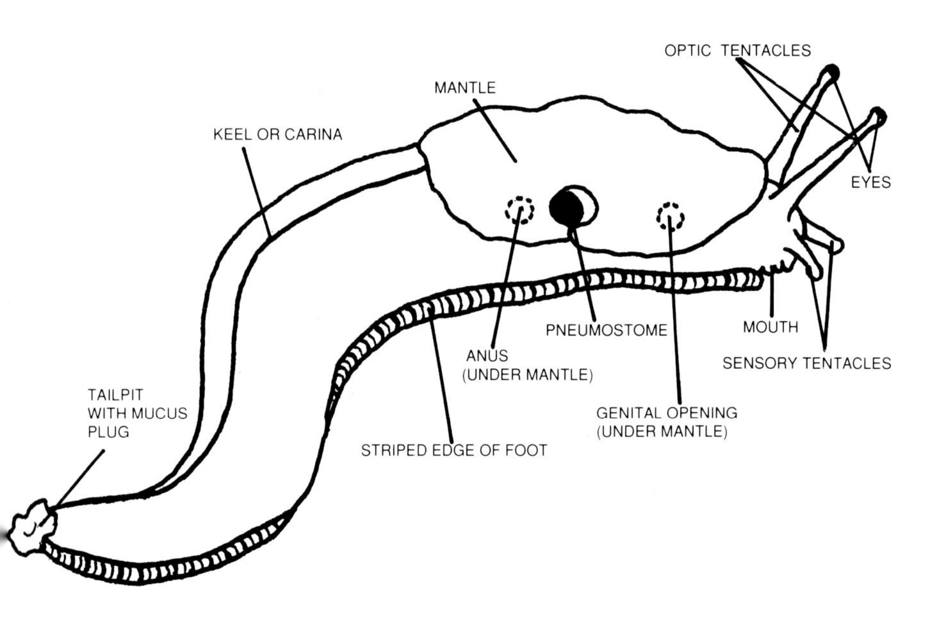

MORPHOLOGY

The slender soft **slug body** functions efficiently in its damp domain. Slugs survive without arms, legs, fur, feathers, or fangs. The smooth, slime-covered body glides along on a muscular **foot**. The foot covers the entire bottom part of its body and the bottom of the foot is referred to as the foot's **sole**. The outer edge or skirt of this foot has dark vertical stripes. The body has a distinctly keeled back, rather like the bottom of a boat. The **keel** or **carina** varies in acuteness with the health of the animal or environmental conditions such as humidity. A **tailpit** contains a mucus plug. Over the front part of the body and head is a head-covering called the **mantle**. The mantle secretes the shell in snails. In slugs, the mantle (where it is exposed) tends to be rather tough and leathery and may provide some protection.

On the right side of the mantle is a hole or slit called the **pneumostome** where the animal breathes. This hole leads to a lung-like cavity. The pneumostome can be opened or closed according to the animal's immediate needs. For example, the slug will open the hole wide when it needs air or close it to keep out rain water. Just forward of the pneumostome, under the flap of the mantle, is the opening of the hermaphroditic slug's male and female **sex organs**. The animal's wastes are expelled from an opening just behind the pneumostome.

There are two sets of **tentacles** on the head. The upper longer pair are the optic tentacles. The actual **eyes** are small black dots at the end of these longer tentacles. The slug eye is not able to perceive detailed images but mainly responds to light intensity. Functioning like a periscope, these eyestems can stretch up and move in all directions to allow the animal a higher and wider view. The tentacles move independently, so the slug can look in two directions at the same time. They can be instantly retracted if the slug senses danger. If a predator does manage to bite off a tentacle, it regenerates. The lower, shorter tentacles are sensory organs for feeling and smelling. Land slugs have a strong sense of smell which directs them to their often odorous food.

The round puckered **mouth** can be protruded with the entire head, out from under the mantle, as it is during the pre-mating stimulation; or can be withdrawn out of sight under the mantle. Inside the mouth is a toothed tongue called the **radula**. The radula is supported by a muscular and cartilaginous projection called the **odontophore**. Over the radula is a platelike jaw that can chop downward like a guillotine. The radula is peculiar only to mollusks and allows them to rasp food into pulp and pull the food into the esophagus. Many tiny teeth, all pointed backwards toward the throat, cover the radula. These filelike teeth are constantly breaking off while new rows of teeth form at the back of the radula.

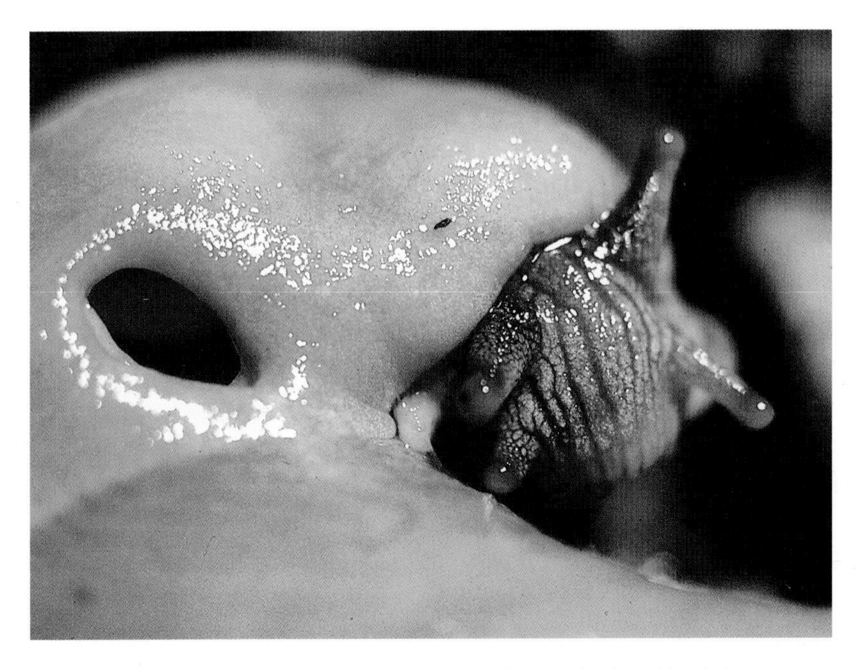

Slug's head showing the two sets of tentacles and the breathing hole.

Various color forms of banana slugs in Washington's Olympic Rain Forest.

COLOR

Land slugs tend to be drably colored creatures. They cannot compare in color to sea slugs (nudibranchia) that are often brilliant colors like electric purple, blue, red, green, orange, or yellow.

Bright yellow banana slugs are exceptional in their color and vividness among land slugs. The cylindrical shape coupled with the banana slug's sometimes butter-yellow hue demonstrates how it received its "banana" name.

However, color does vary in the banana slug. In many areas the slug is spotted against a drab brown, greenish, or golden-brown background color. They can be a solid green-brown or yellow color. They also are sometimes a ghostly candle-wax white while others are dark with spots so close together that they appear to be black.

Surprising changes of color occur in an individual slug with alterations of light, moisture, or food. Color also can change with age, injury, or the health of the animal.

In nature, when an animal is brightly colored, as banana slugs sometimes are, it is often a signal to other animals of poisonous or dangerous qualities. Not all brightly colored creatures are in fact poisonous, but warning colors make predators wary and serve as protection for an animal such as the banana slug.

A form of **camouflage** or **mimicry** does exist for even the bright yellow slugs, though it is hard to believe that they wouldn't be blatantly visible to their common enemies. But many other things exist in these lush green woods that are a similar yellow color, so the banana slug is harder to spot than one might presume. For example- California bay trees (*Umbellaria*), willow trees (*Salix*), and shrubs such as the coffee berry (*Rhamnus*) all drop leaves that in color and shape look very much like a yellow banana slug. The stems of these leaves even resemble the slug's protruding tentacles. Darker colored or spotted banana slugs are well camouflaged and very difficult to see against the background of the forest floor.

A solid yellow banana slug and three bay leaves on the forest floor.

SLIME

Slime (or mucus) is produced by all parts of the body. It plays an important role in the slug's movement, defense, water retention, reproduction, and possibly nutrition.

The pedal gland produces the thick and sticky slime the body travels upon. This slime, plus a suction created by the body, enables the mollusk to stick to slippery surfaces even while traveling upside down. It also acts as a slimy carpet, protecting the slug's soft body from being injured by sharp objects such as broken glass, gravel, or thorns. Another type of slime produced by the foot is thin and watery and moves from the center toward the edges of the foot. A glistening slime trail is left behind instead of footprints.

The slime also protects the animal from being eaten. When the slug is threatened it often assumes a defensive position and emits a very thick mucus. The disturbed slug makes its body short and fat and arches its back with the head and tail partially folded beneath its body. This makes for a slimier and fatter bite than some predators' mouths or beaks can handle. In addition, the thick covering of slime is often repugnant and can cause problems for would-be predators. Dogs and ducks have been seen gagging when they tried to eat a banana slug. Attacking shrews and beetles spend time trying to rid themselves of slime, giving the slug a chance to escape.

The slug's slippery slime and its ability to make its body long and thin, enables it to crawl into extremely small places. This is why they are considered amazing escape artists by people who try to confine them. I have had banana slugs escape through incredibly small holes in a terrarium lid.

The banana slug is able to produce a strong **slime cord** from the mucus plug in the tailpit. Coming face to face with a banana slug slowly lowering itself from a branch by way of a transparent cord can be startling. But this phenomenon is simply a quick and easy way for the slug to return to the ground from a high place.

There are European slugs that mate in midair while hanging from a slime cord.

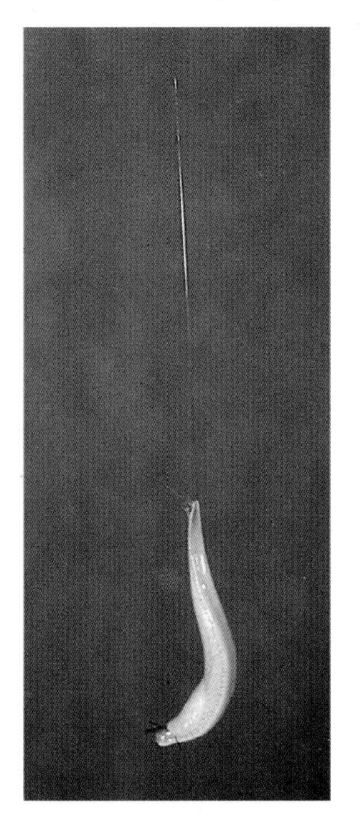

Slug descending via a slime cord.

Spotted banana slug with debris-laden slime at its tail.

Banana slugs prefer to lay down a thick blanket of slime on the ground or on a tree for their marathon mating ritual. During the courtship ritual, prior to mating, the slug partners examine or ingest one anothers' slime.

When debris adheres to the slug's body it is capable of cleaning itself by moving the dirt backwards in the slime until a large hunk of debris-laden slime is at the tail. The slug then turns around and bites off or eats this mass, leaving its body shiny and clean. Since they also are known to eat this mass when it is free of debris, it may serve a nutritional function.

SIZE

The size banana slugs can reach is one of their distinctive characteristics. They are the largest of North American slugs. Most adult *Ariolimax* slugs are about 15-20 cm (six to eight inches) in length. They can, however, reach the length of 25.4 cm (10 inches). In the entire world of land slugs, this size is second only to the huge *Limax cinereoniger* of Europe, which can reach 30 cm (12 inches) in length.

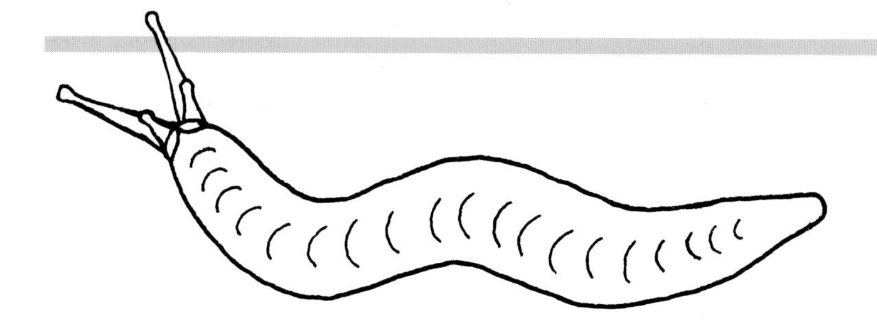

Under view of a slug showing the muscular waves that propel the animal.

LOCOMOTION

Locomotion for slugs is a different process than for other similarly shaped animals, such as caterpillars or worms. Land slugs move by means of locomotor waves on the sole of their foot.

By placing a slug on a piece of glass and viewing it from underneath, half circle waves can be seen moving forward, from tail to head, along the central area of the foot. These waves, or ripples, are muscular contractions lifting up off the surface and then thrusting forward, propelling the slug. The darker waves represent areas that are off the ground.

This type of locomotion does not allow for a quick escape from danger. But a slug's slow tempo of life helps protect it from predators, because it does not catch the eye of larger mammals or birds. Quick movements would be more apt to catch the predator's eye.

Slime is essential for the animal's smooth unobstructed movement. The rim of the foot skids over a carpet of slime secreted by the pedal gland, located just under the head. The slime allows fluidity of movement while at the same time acting as a glue, holding the animal securely to whatever it is crawling upon.

The foot also can clasp slender plant stems as it climbs. Banana slugs are often seen climbing in this manner to feed on small plants and vines.

HOMING - SHELTER SITES

Once the slug finds a suitable place where it can remain moist during dry times, it returns to it for shelter. The shelter sites or "homes" can be a hole in the ground or in the side of a tree, under a log, or places like a crevice in a boulder. During excavation of a forest slope, our neighbor found a banana slug curled up in a hole at the end of a long tunnel, several feet underground. There is a good chance that this slug took over an abandoned tunnel of some other burrowing animal to use as its shelter. Their life is solitary, so these damp homes are not often shared with other slugs.

WATER RETENTION

Land slugs cannot tolerate too much heat or dryness. A slug caught for too long in a hot or dry situation will begin to discolor, visibly shrink and eventually die if it has lost too much of its body's water content. Slugs can readily take up water through the skin so they do not need to take drinks of water but must have a moist environment.

The cruel prank of sprinkling salt on a slug or snail's body has the effect of drawing out the liquids, causing writhing and what appears to be a painful death.

Banana slugs are land dwellers that are thought to be non-swimmers. However, we witnessed a surprising occurrence while walking in the forest after a rain storm. A large banana slug was crawling across a dirt road half covered by a deep puddle of water. The slug crawled, with no hesitation, into the puddle and swam (undulating like a seal) through about 50 cm (20 inches) of deep water. It crawled out the other side, unaffected by being immersed in water for that period of time. This could explain why we have seen banana slugs on large rocks in the middle of streams.

Slug nibbling on a leaf while clasping a plant stalk.

FOOD

What do banana slugs eat? It might be easier to answer the question, "What **don't** banana slugs eat?" Living and decaying vegetation, roots, fruit, seeds, bulbs, lichen, algae, fungi, animal droppings, and carcasses are all eaten by these slugs. Poison oak (*Rhus diversiloba*) is one of the plants they feed upon that is extremely toxic to humans. They are an efficient clean-up crew for the forest floor.

Mushrooms are a favorite food for these slugs and the North American coastal forests have an enormous variety of wild mushrooms. One mushroom book mentions banana slugs grazing peacefully in this fungal jungle. By eating mushrooms and other matter found on the woodland floor, decomposition is hastened while spores and seeds are dispersed through the slug's digestive tract.

Slug eating mushroom gills with an insect on its side.
Gravity has caused the internal shell plate in the mantle to
look like a hump on the slug's back.

Slug eating lichen with its breathing hole opened wide.

Interesting information was uncovered during a research project undertaken in California to find the culprit responsible for damage to coast redwood tree seedlings. A prime suspect was the large native banana slug. The researcher confined banana slugs with tiny coast redwood tree seedlings, but alas, the slugs starved to death or ate their cardboard prison walls rather than eat any part of a *Sequoia sempervirens* seedling. They did, however, eat other tiny plants that sprang up around the seedlings. This eliminated plants that could compete with the redwoods for space, water, light, and nutrients. Furthermore, during this process the banana slugs deposited a valuable nitrogen-rich fertilizer in their droppings, giving young trees a boost in growth. So, banana slugs actually help to enhance survival and growth for the redwood trees that in turn will give them protection from the drying sun. This is a fine example of the interdependence of living things in a balanced natural setting.

When homes are built in the banana slug's native habitat, residents often become upset when they find these animals eating their garden plants. Banana slugs prefer to feed on the wild forest floor, but if humans intrude upon their habitat, most likely they will continue to do some feeding in gardens. I must confess that one year I moved banana slugs to a "relocation center" deeper in the woods when my garden was overpopulated with these yellow creatures. But normally, the native banana slugs that live in my garden do little damage.

Tiny, voracious, garden slug.

NON-NATIVE SLUGS

Not all slugs are as harmless as the native banana slug. There are a number of voracious European slugs that have made their way to North America hidden in cargo or on plants. Having no natural checks on their population, they have spread and multiplied to an alarming degree in urban, rural, and agricultural settings.

Three types of slugs, among the many European ones now established on this continent, have been identified as causing the most serious problems. These are the spotted garden or great gray slug (*Limax maximus*), a small yellow or brown slug (*Arion subfuscus*), and a tiny pale-gray slug (*Agriolimax agrestis*). These shell-less mollusks have munched their way through valuable crops and thousands of gardens.

In the Pacific Northwest another European slug is well established and does extensive plant damage for which the more visible banana slug is often blamed. The slug (*Arion ater*) is a bit thicker and not quite as long as the banana slug. It is solid black or reddish in color and its body has pronounced ridges or furrows.

Besides damaging plants and competing with native mollusks for food, some of these imported slugs attack and often kill other slugs. *Limax maximus* and *Arion subfuscus* have been described as "pugnacious" and are known to attack the much less aggressive banana slug.

Because foreign slugs and snails do so much damage, many people believe all slugs are pests to be eliminated. Banana slugs should not be lumped in with the invaders. For the most part, the banana slug keeps to its native habitat where it is in balance with other plants and animals. If all banana slugs were destroyed, the natural chain of life would be altered in these spectacular western forests.

Furrowed European slug *(Arion ater)* feeding on liverwort and moss.

PREDATORS

Many animals are known to eat banana slugs despite the fact that these slugs are covered by a protective layer of slime. I have seen a Pacific giant salamander (*Dicamptodon ensatus*) and a California newt (*Taricha torosa*) eating a still struggling banana slug and a sow bug (*Oniscus osellea*) feeding on an already dead one. There are reliable accounts of garter snakes, foxes, porcupines, beetles, millipedes, crows, ducks, and other slugs enjoying a meal of banana slugs. One writer watched a raccoon rolling a banana slug on the forest duff so that the slime was covered by debris before eating it. It is also known that raccoons are able to skin amphibians that are covered by a distasteful mucus-like material, before eating them; so they could conceivably do this to slugs as well. Small shrews, shrew moles, and moles, who spend their days underground, have slugs and snails as an important food source. Even though slugs may lay several clutches of eggs a year, their populations are kept well in check by a variety of predators.

It is interesting to note that other animals do not see the same as we do. Nocturnal predators such as the raccoon and fox can not see fine detail and actually perceive green to be a much brighter color than yellow, according to Sandra Sinclair's book <u>How Animals See</u>. This gives the yellow slug a needed advantage against its nighttime enemies.

People also have been known to eat banana slugs. A. Kroeber stated in his <u>Handbook of the Indians of California</u> that the indigenous Yuroks of the North Coast area would use "the large yellow slug" for food when other food became scarce. Another author stated that German immigrant families in the 1800s and early 1900s, ate banana slugs by removing the slime with vinegar, gutting them like fish, and deep frying them in a batter.

A Pacific giant salamander eating a banana slug.

The French are experts at preparing snails for eating. They eat about 45,000 tons of snails every year. Gathered snails are confined for about two weeks and fed on corn meal, grain, or pasta, in order to clean out their digestive systems. This is because land mollusks eat substances that could be poisonous or noxious to humans. If they have eaten plants that have been treated with pesticides it is especially dangerous. People wanting to eat banana slugs should definitely take the trouble necessary to make them a safe meal.

Backlighted banana slug, eating a mushroom, showing mites on its body.

PARASITES

Banana slugs are hosts to a number of parasites that live on the outside or inside of their bodies. A magnified look at a banana slug will often show a large number of tiny mites moving on the body. There is even a type of mite (*Riccardoella limacum*) that is specific to slugs. I have also seen mosquitos on banana slugs, apparently feeding on the slug's colorless blood.

It is known that slugs are intermediate hosts to various parasites that seem to cause little ill effect to the slug, but these parasites can cause serious problems for other animals. According to parasitologists, problems arise when the slug with parasites is ingested by other animals. Tapeworms, flukes, and round worms are some of the parasites that are passed on to animals that eat slugs.

MATING

Pre-mating biting.

Banana slugs are **hermaphrodites**. Each animal has both male and female reproductive organs. There are hermaphroditic mollusks that can fertilize their own eggs, but the banana slug mates with another banana slug. They usually cross fertilize, each producing eggs and sperm simultaneously.

Banana slugs mate at all times of the year. Some researchers believe slugs and snails advertise their readiness to mate by leaving messages for potential partners in their slime trail. Just before mating the banana slugs sometimes eat their partners' slime plug.

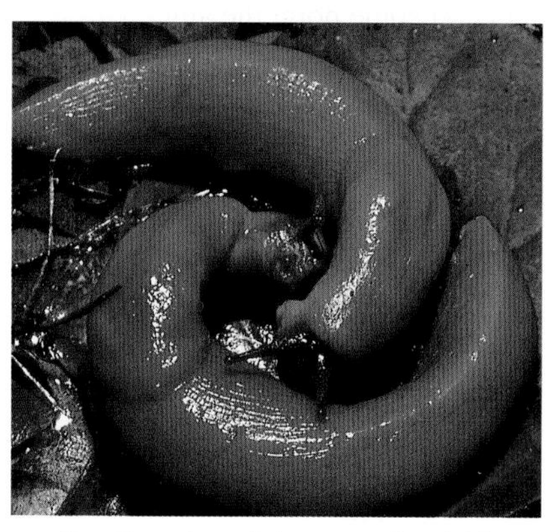

Swelling of genital area on the right side of their heads after hours of stimulation.

The mating ritual starts when two slugs begin circling each other and nudge, lick, and or bite their partner's right side. Some pre-mating biting I have seen becomes quite violent. They often take turns lifting the forepart of their bodies off the ground and striking like a snake with considerable force. During premating stimulation, actual hunks of flesh can be bitten off leaving the right side of each slug battered and scarred.

The slugs are intertwined in an "S" position as they continue to stimulate each other for hours. The genital area just forward of the breathing hole begins to swell as the slugs move closer together. When they are tightly together, penetration and exchange of sperm takes place. They remain pressed together and very still for several more hours.

Double penetration during mating.

When they are ready to separate, difficulties often occur. Because *columbianus*, *dolichophallus* and *californicus* species have evolved such huge male organs, as well as other enlarged internal genital structures, they regularly become stuck and cannot separate. The male organ is often longer than the slug's body. After long hours of premating and mating, more hours are often spent pulling and twisting into incredible positions as they try to pull apart.

At some point the animals finally give up trying to disengage and take turns gnawing off the stuck organ or organs. This paradoxical act of gnawing off the penis is a unique phenomenon known as **apophallation**. Noted malacologists, H. A. Pilsbry and A. R. Mead, stated that the severed male organ possibly regenerates.

My husband and I have observed many banana slug matings and have noted a perplexing detail. Often when the mating is completed, one slug is able to withdraw its male organ into its body while the other slug cannot. So only one male organ is connecting the slugs and must be severed. Perhaps one slug assumes some sort of dominant position which causes its male organ to become anchored.

The size of the genital organs may prove to be only one of the reasons that apophallation has become necessary for these slugs. Apophallation is not completely understood and there

Trying to separate after mating.

One slug beginning to gnaw stuck organ.

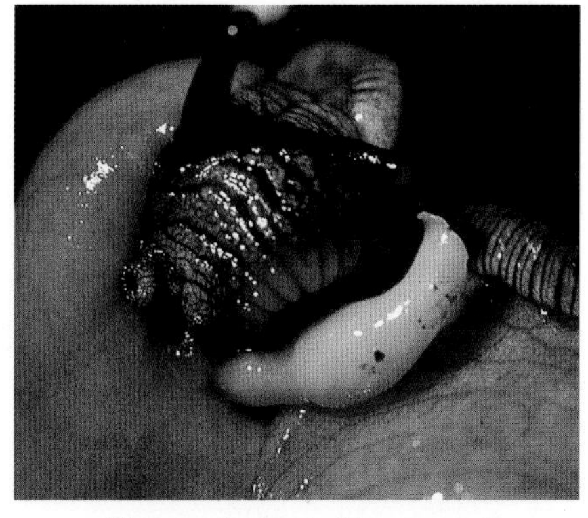

Slug continuing to bite severed male organ after apophallation.

are several intriguing theories that have not yet been proven. There is much more to learn and understand about this remarkable animal.

Albert R. Mead, in his revision of the genus *Ariolimax*, writes about gigantism, which has occurred in this slug's genital structures. He described gigantism as one form of specialization which could and has led to extinction in animals. *Ariolimax* may represent a "blind alley" of sorts, in which the genital structures have evolved to the maximum so that individuals must commit auto-amputation (apophallation) just to maintain the species.

In order to separate, these entangled slugs must apophallate. Free at last, after a mating ritual that may last more than 12 hours, the battered and exhausted slugs crawl away. They can now return to their solitary lives. When conditions are favorable, eggs are laid and new banana slugs begin again their distinctive cycle of life.

Oregon banana slug crawling on lung lichen.

REFERENCES

Becking, R. W. 1967. "Ecology of the Coastal Redwood Forest and the Impact of the 1964 Floods Upon Redwood Vegetation." Final Report, National Science Foundation Grant. GB#4690.

Book of the British Countryside. 1973. Rev. ed. Drive Publications Limited (for the British Automobile Association), London. 536pp.

Fretter, V. and J. Peake (Eds.) 1975. *Pulmonates.* Vol. 1. *Functional Anatomy & Physiology.* New York Academic Press.

Grzimek, B. (Ed.) 1972. *Grzimek's Animal Life Encyclopedia.* Vol. 3. *Mollusks and Echinoderms.* Van Nostrand Reinhold Company, New York. 541pp.

Hochberg, F. G. 1981. "Invertebrate Fauna. 2. Molluscs: Snails and slugs." Vol. 1, p. 53-58 and Vol. 2, p. 203-230. In: C. D. Woodhouse (Ed.). *Literature Review of the Resources of Santa Cruz and Santa Rosa Islands and the Marine Waters of the Channel Islands National Park, California.* Final Technical Report. National Park Service, San Francisco.

Hyman, L. H. 1967. *The Invertebrates.* Vol. 6. *Mollusca.* McGraw-Hill, New York.

Ingram, W. M. and H. M. Adolph. 1943. "Habitat and Observations of *Ariolimax columbianus* Gould." *The Nautilus.* 56: 96-97.

Ingram, W. M. and C. Lotz. 1951. "Land Mollusks of San Francisco Bay Counties." *Journal of Entomology and Zoology.* 42: 5-27.

Ingram, W. M. and A. Peterson. 1947. "Food of the Giant Western Slug *Ariolimax columbianus* (Gould)." *The Nautilus.* 61: 49-51.

Jenkins, M. M. 1972. *The Curious Mollusks.* Holiday House, New York.

Kerney, M. P. and R. A. D. Cameron. 1987. *A Field Guide to the Land Snails of Britain and North-West Europe.* Collins, London. 288pp.

Kozloff, E. N. 1978. *Plants and Animals of the Pacific Northwest.* University of Washington Press, Seattle. 264pp.

Kroeber, A. L. 1925. *"Handbook of the Indians of California." Bureau of American Ethnology Bulletin 78.* Washington, D.C. (See p. 84.)

Larson, A. W. 1963. "The Banana Slug." *Pacific Discovery Magazine.* September-October, p. 10-11.

Mead, A. R. 1943. "Revision of the Giant West Coast Land Slug of the Genus Ariolimax Moerch (Pulmonata: Arionidae)." *The American Midland Naturalist.* 30: 675-717.

Pilsbry, H. A. 1939-48. *Land Mollusca of North America.* Monograph 3, 2 vols. Academy of Natural Sciences, Philadelphia. (See p. 710.)

Richter, K. O. 1980. "Movement, Reproduction, Defense, and Nutrition as Functions of the Caudal Mucus in *Ariolimax columbianus.*" *The Veliger.* 23: 43-47.

Rollo, C. D. and W. G. Wellington. 1977. "Why Slugs Squabble." *Natural History Magazine.* November, p. 46-51.

Runham, N. W. and P. J. Hunter. 1970. *Terrestrial Slugs.* Hutchinson University Library, London.

Sinclair, S. 1986. *How Animals See: Other Visions of Our World.* Facts on File, New York.

Solem, A. 1974. *The Shell Makers: Introducing Mollusks.* J. Wiley and Sons, New York.

Waters, G. 1976. "The Beginners Guide to Slugs and Snails." *Pacific Horticulture Magazine.* April.

INDEX

Bay Leaves Press
160 Robideaux Road
Aptos, Ca 95003 U.S.A.